LEXINGTON PUBLIC LIBRARY

Ankylosaurus
Lori Dittmer

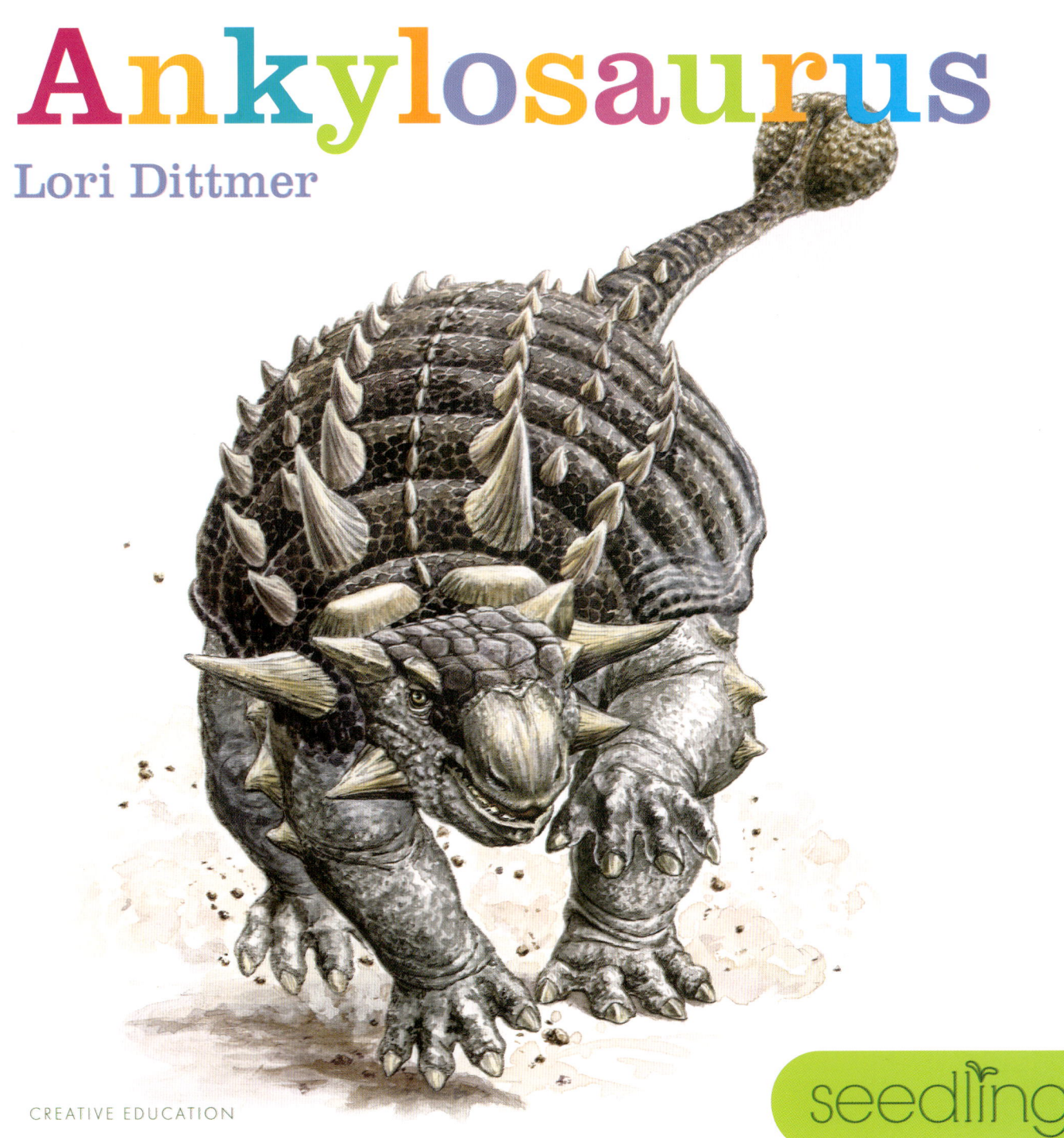

CREATIVE EDUCATION
CREATIVE PAPERBACKS

seedlings

Published by Creative Education and Creative Paperbacks
P.O. Box 227, Mankato, Minnesota 56002
Creative Education and Creative Paperbacks
are imprints of The Creative Company
www.thecreativecompany.us

Design by Ellen Huber
Production by Rachel Klimpel and Ciara Beitlich
Art direction by Rita Marshall

Photographs by Alamy (Daniel Eskridge, NAPA, Universal Images Group North America LLC / DeAgostini), iStock (Ieonello), National Geographic (Royal Tyrrell Museum), Science Source (Deagostini, James Kuether/ Science Photo Library, SPL), Shutterstock (Catmando, Daniel Eskridge, Herschel Hoffmeyer, Ralf Juergen Kraft, Warpaint), SuperStock (Stocktrek Images), Tumblr (American Museum of Natural History)

Copyright © 2024 Creative Education, Creative Paperbacks
International copyright reserved in all countries.
No part of this book may be reproduced in any form
without written permission from the publisher.

Library of Congress Cataloging-in-Publication Data
Names: Dittmer, Lori, author.
Title: Ankylosaurus / by Lori Dittmer.
Description: Mankato, Minnesota : Creative Education and Creative Paperbacks, [2024] | Series: Seedlings: dinosaurs | Includes bibliographical references and index. | Audience: Ages 4–7 | Audience: Grades K–1 | Summary: "Early readers are introduced to Ankylosaurus, a Cretaceous plant eater. Friendly text and dynamic photos share the dinosaur's looks, behaviors, and diet, based on scientific research"— Provided by publisher.
Identifiers: LCCN 2022013862 (print) | LCCN 2022013863 (ebook) | ISBN 9781640265004 (library binding) | ISBN 9781682770528 (paperback) | ISBN 9781640006300 (ebook)
Subjects: LCSH: Ankylosaurus—Juvenile literature. | Dinosaurs—Juvenile literature.
Classification: LCC QE862.O65 D583 2024 (print) | LCC QE862.O65 (ebook) | DDC 567.915—dc23/eng20221202
LC record available at https://lccn.loc.gov/2022013862
LC ebook record available at https://lccn.loc.gov/2022013863

Printed in China

TABLE OF CONTENTS

Hello, *Ankylosaurus*! 4

Cretaceous Dinosaurs 6

Early Discovery 8

Living Tanks 10

Tail Club 12

Plant Eaters 14

What Did *Ankylosaurus* Do? 16

Goodbye, *Ankylosaurus*! 18

Picture an *Ankylosaurus* 20

Words to Know 22

Read More 23

Websites 23

Index 24

Hello, Ankylosaurus!

This dinosaur lived long ago.

Tyrannosaurus rex and *Triceratops* lived then, too.

Barnum Brown named it in 1908.

Ankylosaurus means "stiff lizard." Its back bones grew together.

Ankylosaurus was about the weight of an elephant.

It moved on four legs. Bony plates and spikes covered its head and back. Its skin was hard.

The tail ended in a bony club. *Ankylosaurus* might have used it for fighting.

The club could break bones.

Ankylosaurus ate plants.

It grabbed leaves in its strong beak.

Heavy *Ankylosaurus* swung its horned head.

Food was near!

It sniffed the air.

Goodbye, Ankylosaurus!

Picture an *Ankylosaurus*

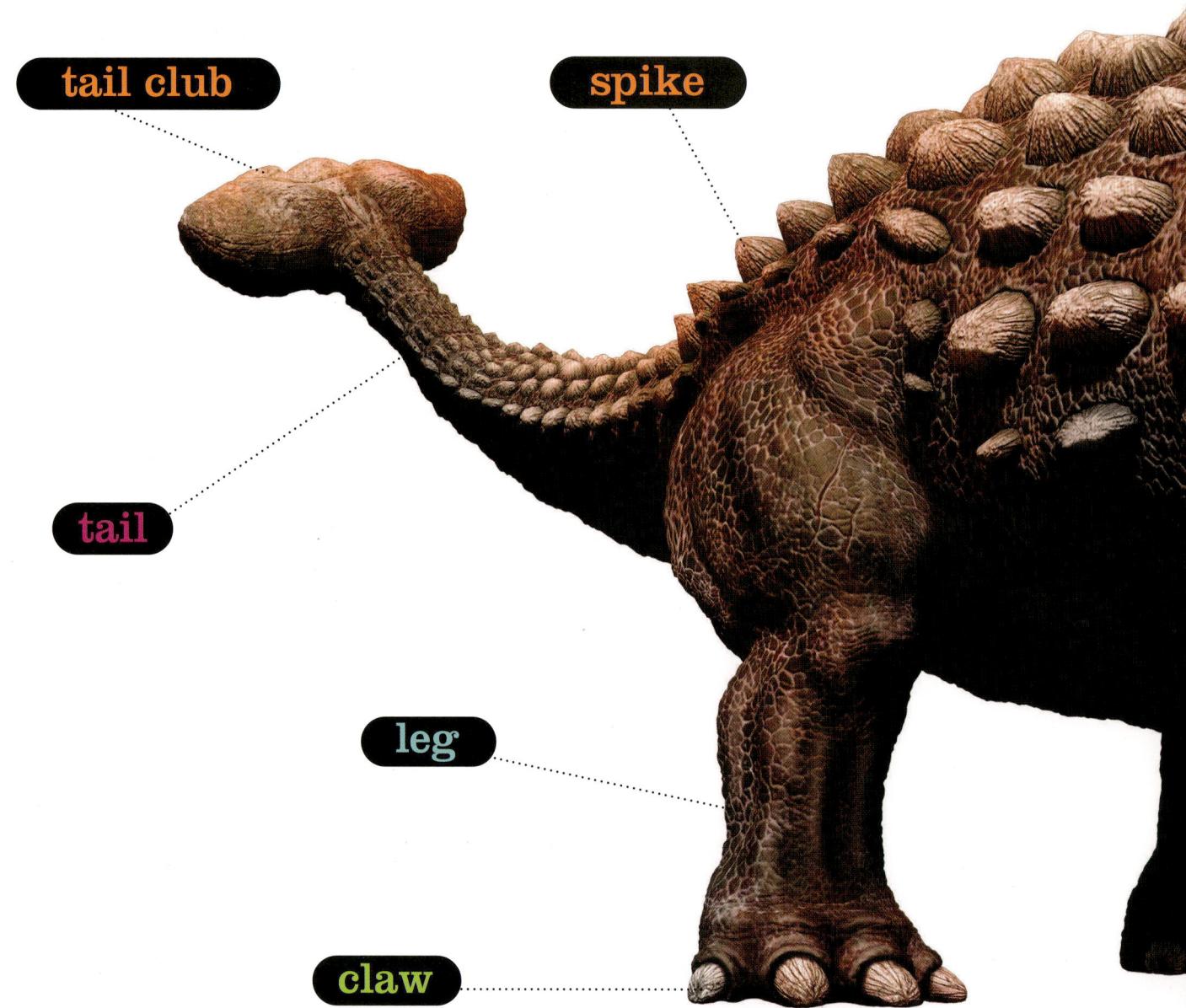

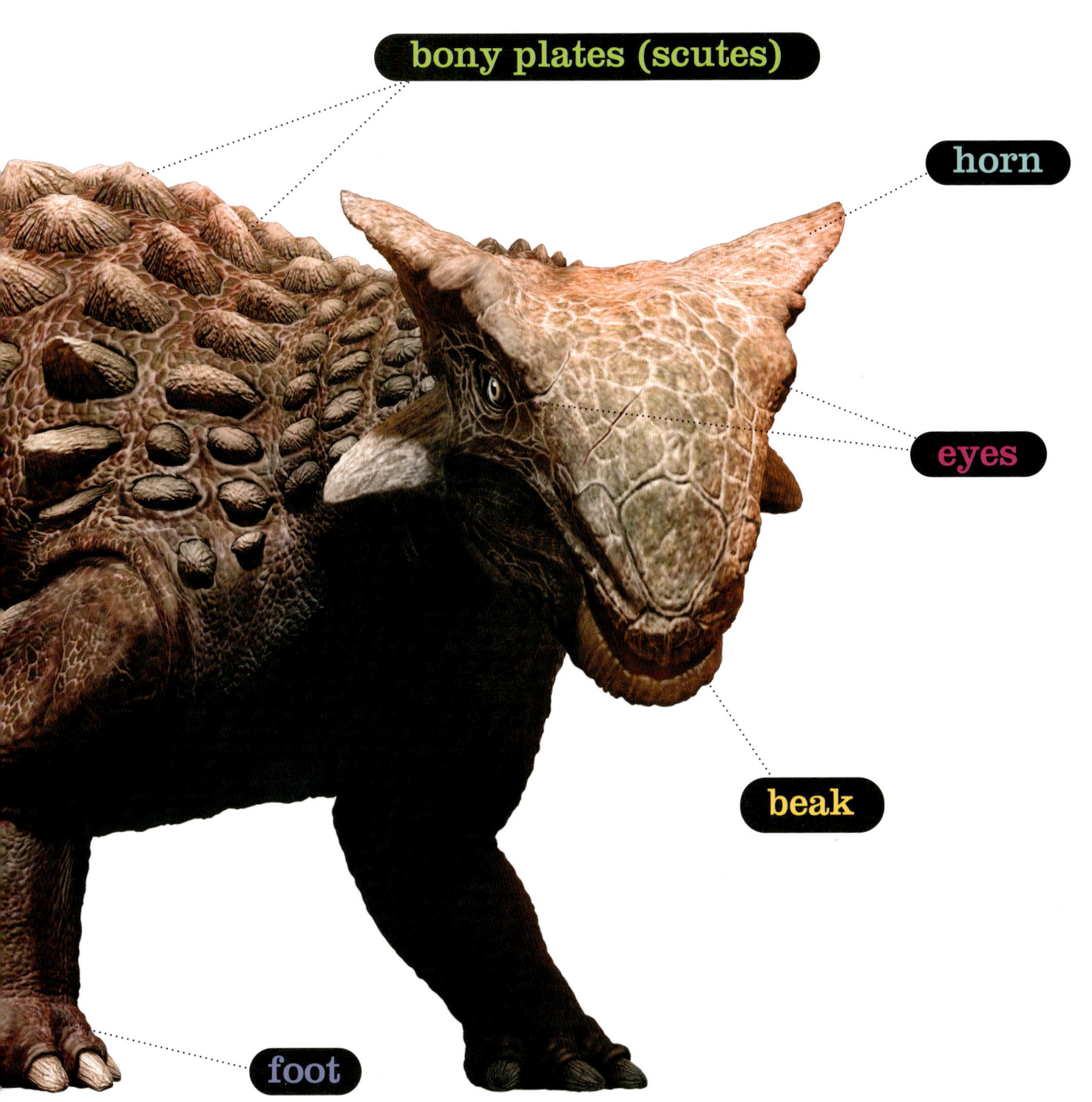

Words to Know

beak: the hard, pointed part of the mouth of some animals

club: the bony part at the end of a tail

spike: a sharp, pointed object

Read More

Pimentel, Annette Bay. *Do You Really Want to Meet Ankylosaurus?* Mankato, Minn.: Amicus, 2020.

Sabelko, Rebecca. *Ankylosaurus.* Minneapolis: Bellwether Media, 2020.

Websites

DK Find Out! | *Ankylosaurus*
https://www.dkfindout.com/us/dinosaurs-and-prehistoric-life/dinosaurs/ankylosaurus
See a picture of what this dinosaur might have looked like and take a dinosaur quiz.

PBS Kids | Dinosaur Discoveries: *Ankylosaurus*
https://pbskids.org/video/dinosaur-train/1346704087
Watch a video about *Ankylosaurus.*

Note: Every effort has been made to ensure that the websites listed above are suitable for children, that they have educational value, and that they contain no inappropriate material. However, because of the nature of the Internet, it is impossible to guarantee that these sites will remain active indefinitely or that their contents will not be altered.

Index

discovery, 8
feeding, 14, 15, 17
fighting, 12, 13
head, 11, 16
name, 8, 9
plates and spikes, 11
skin, 11
tail club, 12, 13
weight, 10
when it lived, 6, 7